The Life and Times of the PEANUT

— of the —

Charles Micucci

Houghton Mifflin Company   Boston

For all the little goobers in the peanut gallery

The text of this book is set in 15-point Plantin.
The illustrations are watercolor, reproduced in full color.

Library of Congress Cataloging-in-Publication Data
Micucci, Charles.
Life and times of the peanut / Charles Micucci.
p. cm.
Summary: Examines the history and statistics of peanuts, their agriculture and influence.
RNF ISBN 0-395-72289-6   PAP ISBN 0-618-03314-9
1. Peanuts — Juvenile literature. [1. Peanuts.] I. Title.
SB351.P3M535   1997   96-1290   641.3'56596 — dc20   CIP   AC

Manufactured in China
LEO 20 19 18 17 16 15 14 13 12 11
4500302230

For Further Reading:

The Great American Peanut: An Overview of the U.S. Peanut Industry. Alexandria, Va.: National Peanut Council, 1995.
Holt, Rackham. *George Washington Carver.* Garden City, N.Y.: Doubleday, Doran and Co., Inc., 1945.
Johnson, Frank Roy. *The Peanut Story.* Murfreesboro, N.C.: Johnson Publishing Co., 1964.
Pattee, Harold, and Clyde Young. *Peanut Science and Technology.* Yoakum, Tex.: American Peanut Research and Education Society, 1982.
Peanut Trivia. Tifton, Ga.: Georgia Peanut Commission.
Royal Botanical Gardens, KEW. *Ground-nut or Pea-nut.* Great Britain, 1901.
Steward, Julian H., ed. *Handbook of South American Indians.* Washington, D.C.: Smithsonian Institution, Government Printing Office, 1946.
U.S. Department of Agriculture. *Growing Peanuts.* Washington, D.C.: Government Printing Office, 1954.
U.S. Department of Agriculture. *Peanuts: Culture and Uses.* Washington, D.C.: Government Printing Office, 1889.
Woodroof, Jasper G., ed. *Peanuts: Production, Processing, Products.* Westport, Conn.: Avi Publishing Co., 1983.

Acknowledgments:

The author expresses his gratitude to the Planters and LifeSavers Companies of Nabisco for permission to use Mr. Peanut on page 29. Mr. Peanut is a registered trademark of Nabisco Inc.
"Goober Peas" (words and music) was originally published by A. E. Blackmar, New Orleans, 1866. It is believed that anonymous Confederate soldiers wrote the main melody and verses, and that A. E. Blackmar may have completed the song under the pseudonyms of A. Pindar and P. Nutt.

Contents

The Humble Peanut

Peanuts are one of our most beloved yet least understood crops. Through the years, they have been called goobers, pinders, ground peas, and many names in between.

Peanuts were first grown thousands of years ago by South American Indians and were later transported to other cultures. Today, people all over the world enjoy peanuts.

The peanut first grew in South America about 5,000 years ago. Potatoes, cocoa, lima beans, and pineapples also originated in South America.

peanuts   potatoes   cocoa   lima beans   pineapples

Peanuts are one of the world's oldest crops. South American Indians were growing peanuts before Romans were planting olive trees.

South America 1500 B.C.   Rome 600 B.C.

In the 1500s, Spanish and Portuguese explorers shipped peanuts from South America to Asia, Europe, and Africa.

Nowadays, people in China, Japan, and Singapore cook peanuts with rice and other foods. In Indonesia and India, peanut oil is used to light lamps, while in America, peanuts are served at public events such as baseball games, circuses, and county fairs.

America's Favorite Nut?

Each year Americans eat more peanuts than almonds, hazelnuts, and walnuts put together. The peanut is so popular that many people claim it as their favorite nut. There's just one problem . . . the peanut is not a nut, it is a legume.

Legumes are flowering plants that bear seeds in pods and have round growths on their roots called nodules. Tiny plants that live in the nodules help legumes by making nutrients similar to fertilizer. These nutrients travel from the roots to all parts of the plant so it can grow strong. There are more than 10,000 kinds of legumes, including peas, beans, clover, and alfalfa.

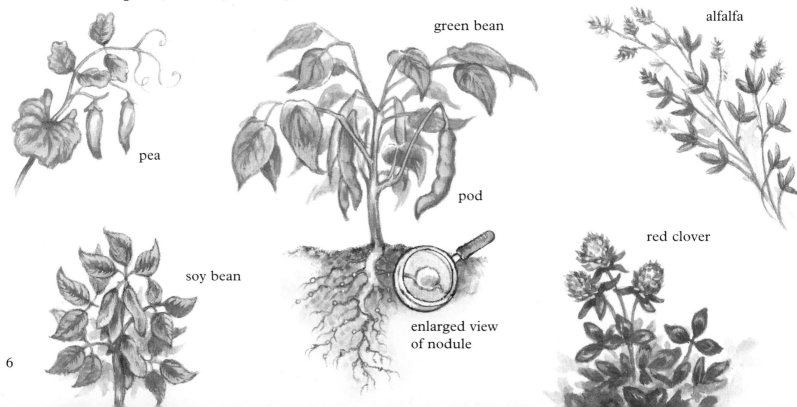

pea

green bean

alfalfa

soy bean

pod

red clover

enlarged view
of nodule

The Uniqueness of Being a Peanut

A peanut doesn't grow in a thick, hard shell like a walnut. It grows in a pod like a bean.

But the peanut pod doesn't grow above ground like the bean pod. It grows below ground.

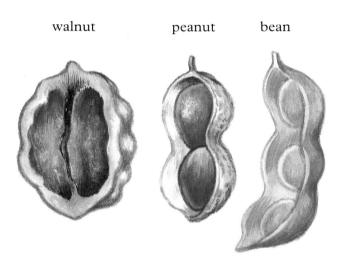

walnut   peanut   bean

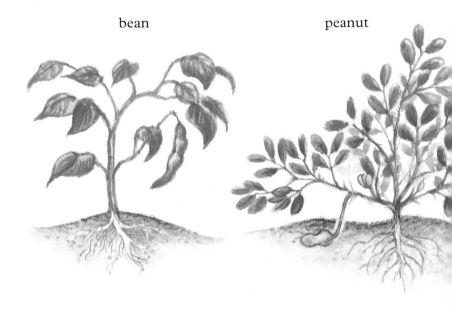

bean   peanut

And though they are both underground, peanuts don't grow like potatoes, which are part of underground stems.

Nor do they grow like carrots, which are part of underground roots.

Peanuts are shoots from flowers that blossom above ground, but produce seeds — peanuts — below ground.

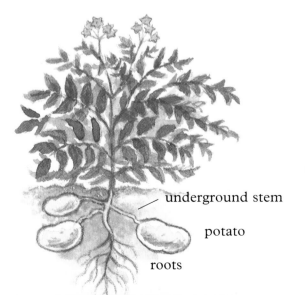

underground stem

potato

roots

carrot

roots

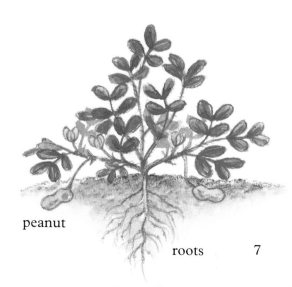

peanut

roots

7

The Amazing Peanut Plant

When you eat a peanut, you are actually eating the seed of the peanut plant. Under the right conditions, a peanut seed will grow into a plant about eighteen to twenty-four inches tall.

A peanut plant has fuzzy stems, oval leaflets, and little yellow flowers shaped like butterflies. But it's below ground where the plant's treasures lie.

There are two types of peanut plants: runners, which spread out like vines, and bunch peanut plants, which grow upright like small bushes.

Runner peanuts grow beneath
the plant's branches.

Bunch peanuts grow close to the
plant's main stem and tap root.

Parts of a Peanut Plant

Leaflets grow as pairs in groups of four.

The **main stem** supports the plant.

Stems carry water and food to the leaflets.

Peanut **leaflets** fold up in pairs at night.

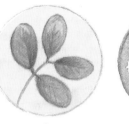

day   *night*

Yellow **flowers** about 1/2 inch across are low on the plant.

Most peanut **flowers** bloom for one day and then wilt.

The **primary branches** are large stems where the peanut flowers blossom.

Pegs are shoots that sprout from the base of the wilted flowers. They dive into the ground and their dark tips swell into peanuts.

The **tap root** is the main root that holds the plant in the ground.

A peanut **pod**, or shell, has a ridged surface and forms underground.

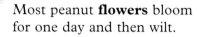

Roots branch off from the tap root and soak up water and food from the soil.

Nodules on the roots make more food to help the plant grow strong.

The pod usually holds two **seeds** (the peanuts) wrapped in protective skin.

peanut with skin removed

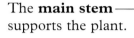

How Peanuts Grow

Peanut plants grow best in sandy or loose soil with warm, sunny weather and moderate rain. Most peanuts need about five months to grow to maturity.

2 inches deep

About one to two weeks after a peanut is planted, a bud breaks through the ground.

For the next two months, the plant grows many leaflets, which absorb energy from the sun.

During summer, little yellow flowers blossom. They open at sunrise.

The next day the flower wilts. A few days later the base of the flower's stem starts to grow.

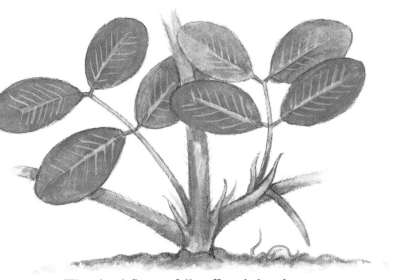

The dead flower falls off and the shoot, called a peg, grows toward the ground.

A few weeks later, the peg enters the ground and the tip turns horizontal.

By late summer, the tip has grown into a peanut shape, and its outer shell hardens.

About five months after planting, the new peanuts are ready for harvest.

Most peanut plants in gardens produce about twenty peanuts. But farmers who use special seed peanuts may harvest more than one hundred peanuts per plant.

Harvesting Peanuts

In small gardens, peanut plants are dug up with shovels and the peanuts picked by hand. On large farms, special equipment is used to harvest peanuts.

First the farmer takes the peanut plants out of the ground and lays them in rows called windrows. His tractor has three attachments: a digger, a shaker, and an inverter.

The **inverter** turns the plant upside down so the peanuts can dry in the sun.

The **digger** plows through the soil about six inches below the peanut plant and cuts the tap root.

The **shaker** scoops up the plant and shakes the dirt off the peanuts.

After the plants have dried for at least two days, a combine gathers up the windrows.

Inside the combine the peanut plant is spun around, separating the peanut from the plant.

Then the peanuts are blown through a chute into a screen mesh hopper.

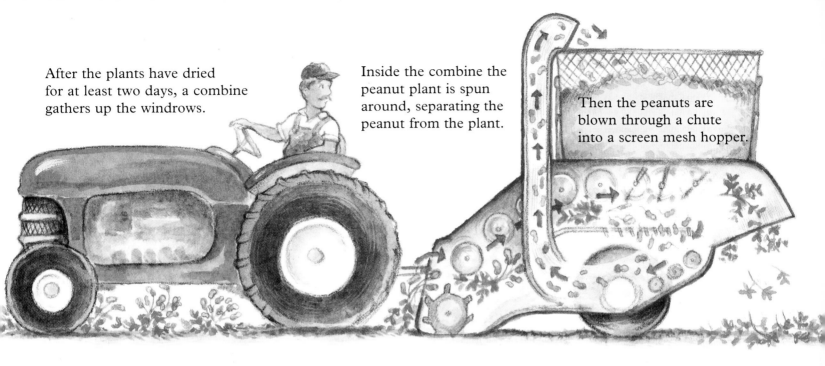

When the hopper is full, the farmer empties the peanuts into a special drying wagon.

Warm air is pumped through the floor to dry the peanuts. Drying peanuts is important because fresh-dug peanuts are 25 to 50% moisture, which has to be reduced to 10% or the peanuts may get moldy.

Warm air goes in here.

Finally, the farmer takes the peanuts to a buying station where they are inspected for quality and sorted by size: larger peanuts are used for dry-roasted peanuts; smaller peanuts are used in candies, peanut butter, and other products.

Tons of Peanuts

Worldwide, farmers harvest more than 26 million tons of peanuts a year. The largest peanut-producing countries are China, India, and the United States. Together they raise more than 90% of the world's peanuts. In the United States, the greatest peanut-producing states are Georgia, Texas, North Carolina, and Alabama. Georgia peanut farmers grow 44% of all American peanuts.

Twenty-six million tons of peanuts equals 9 pounds of peanuts for every person on Earth.

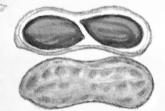

The main peanut-growing areas in the United States are shaded with peanuts.

In the United States, peanut farmers usually grow one of four peanut varieties

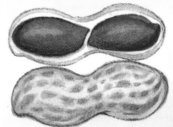

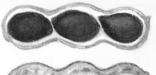

Runner Peanuts

Grown in Georgia, Alabama, Florida, Texas, and Oklahoma, these peanuts account for more than half of all peanuts grown in the U.S.

Virginia Peanuts

Grown in Virginia and North Carolina, these have larger seeds than other peanuts.

Spanish Peanuts

Grown in Texas and Oklahoma, these contain more peanut oil than other peanuts.

Valencia Peanuts

Grown in New Mexico and Tennessee, these can have three or more peanuts per pod.

Leading Peanut-Growing States

If the state of Georgia were a separate country, it would be the fifth largest peanut-growing country in the world.

South Carolina | New Mexico | Florida | Oklahoma | Virginia | Alabama | North Carolina | Texas | Georgia

Leading Peanut-Growing Countries

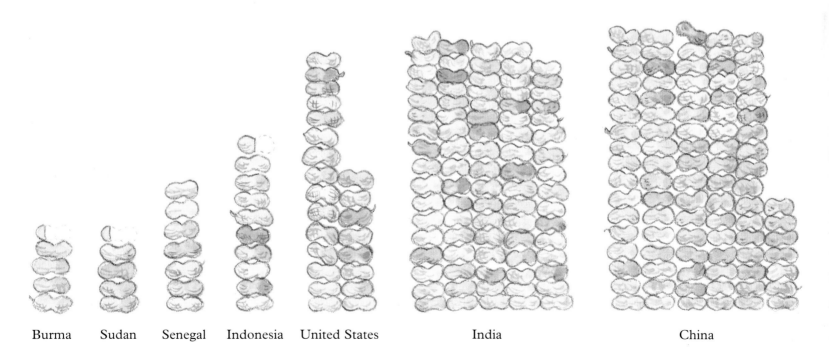

Burma | Sudan | Senegal | Indonesia | United States | India | China

Each peanut symbol equals 100,000 tons. (1994 figures)
National Agricultural Statistics Services and Foreign Agricultural Service

The Peanut Can Do It All

Peanuts are one of the world's most beneficial crops. Every part of the plant has a use, but no part is as versatile as the peanut itself, which is eaten alone, added to many other foods, squeezed into peanut oil, and processed into a variety of products such as soap and explosives. In America most peanuts are grown for food, but in many other countries peanuts are raised primarily for peanut oil.

Uses for peanuts as food:

Peanut seeds are 26% protein, 46% fat (peanut oil),
high in energy, and contain B vitamins.

roasted or boiled peanuts

salted peanuts

chocolate-covered peanuts

peanut butter

peanut ice cream

peanut brittle

cookies

peanut milk

salad dressing

peanut flour

imitation cheese

margarine

Chefs like to cook with peanut oil because it can be heated to a high temperature than most oils and it doesn't easily absorb odors.

Uses for peanuts in industry:

facial creams

paints

shampoo

lamp oil

soap

textile fibers

high-protein
livestock food

explosives

lubricating machines

Uses for peanut pods (shells):

Uses for stems and leaflets:

fire logs

cat litter

wall boards

hay for livestock

green manure
(fertilizer)

Use for peanut skins:

Use for roots:

paper

plowed under to enrich soil

smooth peanut butter

chunky peanut butter

natural peanut butter — peanut oil — peanut butter

From Peanuts to Peanut Butter

More than half of all peanuts in America are ground into peanut butter. When peanuts are ground, oil in the peanut is released and mixes with the fleshy part of the peanut to create a soft, butterlike paste. The most popular type of peanut butter is smooth, and then chunky, which has peanut pieces added to it.

Most food companies add stabilizers such as vegetable oil to peanut butter to make it easier to spread and to keep the oil from oozing out of the peanut butter. Some people prefer natural peanut butter because it has no stabilizers. It's usually harder to spread and has a layer of peanut oil on top.

A large peanut butter company can produce over 10,000 pounds of peanut butter an hour. That is the same weight as a full-grown elephant.

U.S. food law requires that peanut butter must contain at least 90% peanuts. The other ingredients can be salt, sweeteners, and stabilizers.

How Peanut Butter Is Made

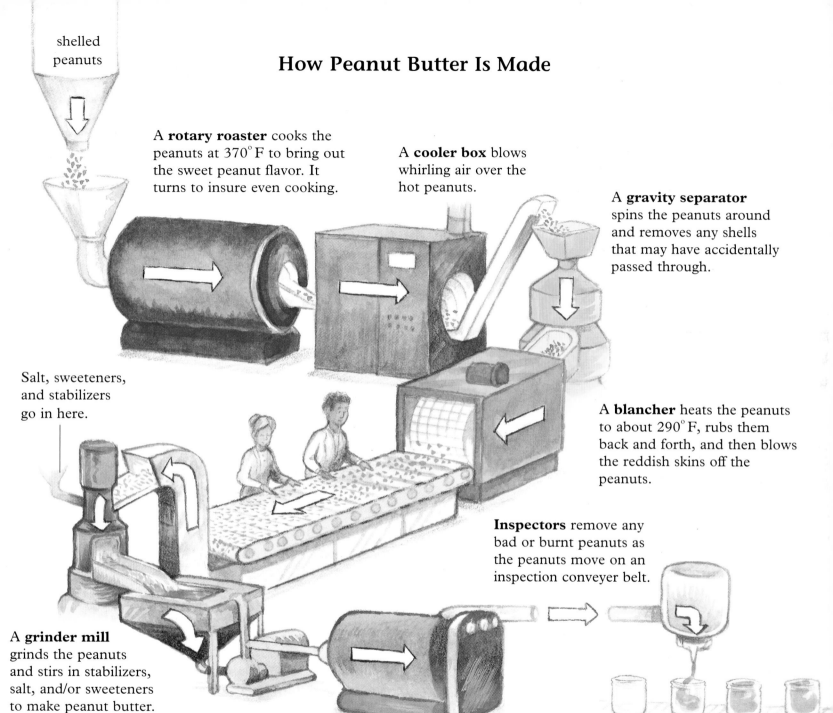

shelled peanuts

A **rotary roaster** cooks the peanuts at 370° F to bring out the sweet peanut flavor. It turns to insure even cooking.

A **cooler box** blows whirling air over the hot peanuts.

A **gravity separator** spins the peanuts around and removes any shells that may have accidentally passed through.

Salt, sweeteners, and stabilizers go in here.

A **blancher** heats the peanuts to about 290° F, rubs them back and forth, and then blows the reddish skins off the peanuts.

Inspectors remove any bad or burnt peanuts as the peanuts move on an inspection conveyer belt.

A **grinder mill** grinds the peanuts and stirs in stabilizers, salt, and/or sweeteners to make peanut butter.

A **homogenizer** stirs the peanut butter smoothly and pumps it through pipes.

A **votator** — a revolving refrigerator — cools the peanut butter. This traps the peanut butter and stabilizers together so they don't separate later.

Finally, peanut butter is pumped into jars.

900 B.C   1700   1890   1904   Today

Three Thousand Years of Peanut Butter

Each year, Americans eat more than 800 million pounds of peanut butter. Most of it is eaten by children. But American children were not the first peanut butter eaters.

Almost 3,000 years ago, South American Indians ground peanuts into a sticky paste. Their peanut butter was not as spreadable as modern peanut butter and it tasted different, too. They mixed their gooey delight with cocoa, the main ingredient of chocolate.

Modern peanut butter, which is easier to spread, was invented in 1890 by a doctor in St. Louis. By 1900 the store shelves in many parts of America were stocked with peanut butter.

900 B.C.
South American Indians ground peanuts into paste and mixed it with cocoa.

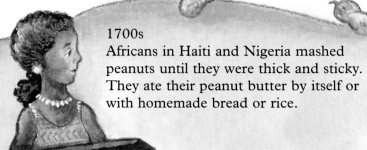

1700s
Africans in Haiti and Nigeria mashed peanuts until they were thick and sticky. They ate their peanut butter by itself or with homemade bread or rice.

1904
Peanut butter was served at the St. Louis World's Fair.

A.D. 1500
The Aztecs in Mexico soothed aching gums with peanut paste. They used many plant products for medicine.

1890
A St. Louis doctor fed peanut butter to his elderly patients because it contained almost as much protein as meat yet was easier to digest.

A peanut butter sandwich and an 8 oz. glass of milk supplies an 8- to 10-year-old child with 63% of their daily protein requirement.

The average American child eats more than 1,500 peanut butter sandwiches by the time they graduate from high school. That's a stack of sandwiches more than 125 feet tall.

Americans eat more peanut butter than any other people in the world.

It takes 720 peanuts to make a one-pound jar of peanut butter.

GOSH, IT'S CROWDED IN HERE.

1920s–30s
Mothers across America made peanut butter sandwiches, which provided low-cost protein and energy, for their children.

1968–1972
During the Apollo space flights, dry 3/4" peanut butter sandwiches were sent into space. When an astronaut chewed the mini sandwich his saliva added moisture to it.

1943
At the height of World War II, the United States Army fed its soldiers 57 million pounds of peanut butter.

1992
The National Peanut Council sent 30 tons of peanut butter, enough for 500,000 sandwiches, to Russia.

The First Peanut Farmers

South American Indians were growing peanuts more than 3,500 years ago. At first they gathered wild peanuts. Later, they grew their own peanuts from seed. As tribes shared their cultures, the peanut spread across South America.

The largest group of peanut farmers were the Incas, who lived in Peru and Ecuador from about A.D. 1200 to 1532. It is believed that the Incas offered peanuts, which they called ynchic, to the sun during religious ceremonies.

Incas worked together on community farms and used llamas to transport their crops. In addition to peanuts, they also grew beans, corn, cotton, papayas, peppers, potatoes, pineapples, and sweet potatoes.

During the dry season and droughts, the Incas brought water to their fields by digging ditches that drained water from lakes and streams.

South American Indians grew peanuts in a variety of places

Incas from Peru
valleys between mountains

Mojo and Bauré from Bolivia
sandy beaches along lakes and rivers

Various tribes from Brazil
jungle clearings

The Ancón people who lived on the coast of Peru from 500 to 750 B.C. buried their ancestors with peanuts and other foods so they wouldn't be hungry in the afterlife.

The Moche people, who lived in Peru from 200 B.C. to A.D. 800 decorated pottery with peanut shells.

Among many Brazil Indian tribes, the women planted and harvested peanuts. It was believed that if men cared for the plants, they wouldn't yield any peanuts.

Eventually South American Indians traded peanuts with the peoples of Central America, Mexico, and the Caribbean Islands. But the peanut's travels were just beginning.

Sailing Across the Ocean

In the 1500s, Spanish explorers conquered the Incas, and Portuguese explorers defeated many of the Brazil tribes. After their conquests the explorers shipped peanuts to Europe, Asia, and Africa.

The peanut's arrival in the United States is a mystery, but African slaves are credited with developing its culture. They planted peanuts in their personal gardens and inspired other farmers to grow peanuts.

AFRICA

Congo

Zaire

Angola

Some slaves called peanuts *goobers*, after the Kimbundu word for peanut, *nguba*. Today many southerners still call peanuts goobers.

Green area of map: Kimbundu language.

In South Carolina, peanuts are nicknamed *pinders*, sometimes spelled *pindars*, from the Kongo word *mpinda*. Peanuts were first exported from South Carolina to other states after the Revolutionary War.

Red area of map: Kongo language.

In the late 1700s and early 1800s, free Africans planted peanuts on their farms and were the first to sell them in markets.

Peanut Travels

British scientists, amazed by peanuts because they grew underground, called them *earth nuts* or *ground nuts*.

NORTH AMERICA

The first peanut plantations in America grew peanuts to feed chickens, turkeys, and pigs. Before the Civil War, plantation owners called peanuts ground peas.

EUROPE

During the winter of 1830, when French olive trees were damaged, peanut oil was sent from Gambia in Africa to replace olive oil.

Chocolate-covered peanuts have been a treat in Spain since the early 1800s.

In 1602, Captain Gosnold saw "groundnuts" growing near the Virginia coast. In 1607 he piloted one of the three ships that took colonists to Jamestown.

Indians of the Caribbean Islands grew peanuts, which they called *mani*, in their gardens.

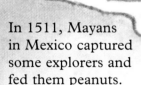

In 1511, Mayans in Mexico captured some explorers and fed them peanuts.

By the 1800s, peanuts were grown throughout Africa. Many African peanuts were crushed into oil or made into soap and sent to Europe.

AFRICA

Hundreds of years ago in Africa, peanuts were cast in bronze and gold. Some Africans believed that peanuts and other plants had souls.

SOUTH AMERICA

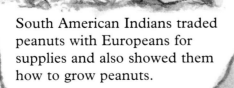

South American Indians traded peanuts with Europeans for supplies and also showed them how to grow peanuts.

PACIFIC OCEAN

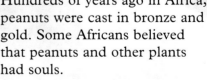

ATLANTIC OCEAN

25

Goober Peas

In the early days of America, only a few people liked peanuts. That changed during the Civil War (1861–1865), when soldiers from the South and North ate peanuts, then known as goober peas. Peanuts were a popular food with the troops because they could be easily transported in a pouch or backpack.

When in camp, the troops could roast them on a campfire or boil them in water. When on the run, soldiers ate peanuts raw. Near the end of the war, when food was scarce, many Southern soldiers ate nothing except goober peas.

Peanut oil was used by southern railroad engineers to lubricate their locomotives, while southern women cooked with it instead of lard and butter.

Many Northern soldiers saw peanuts for the first time when they marched past peanut fields in Virginia.

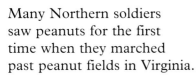

During the war, Confederate soldiers sang this song, called "Goober Peas," about peanuts.

Words by A. Pindar, Esq.　Music by P. Nutt, Esq.

1. Sit- ting by the road side　on　a sum-mer day,　　Chat-ting with my mess mates
2. When a horse man pass-es, the soldiers have a　rule ,　To cry out　at　their loud- est

pass- ing time a- way,　　Ly-ing in the　shadow　　un-der-neath the trees,
"Mis- ter here's your mule,"　　But an-oth-er　pleasure　en-chant-ing-er　than these, Is

Chorus

Good-ness how de- li- cious,　eating goober peas!　　Peas! Peas! Peas! Peas!
wearing　out your Grinders,　eating goober peas!

eat- ing goo- ber peas!　Goodness how de- li- cious, eating goober　peas!

3. *Just before the battle the General hears a row,*
He says "the Yanks are coming, I hear their rifles now,"
He turns around in wonder, and what do you think he sees
The Georgia Militia, eating goober peas!
chorus

4. *I think my song has lasted almost long enough,*
The subject's interesting, but rhymes are mighty rough,
I wish this war was over when free from rags and fleas,
We'd kiss our wives and sweethearts and gobble goober peas!
chorus

The Peanut Saves the Farm

After the Civil War, returning soldiers told their families about peanuts. Soon vendors were selling peanuts at public events such as circuses and baseball games.

But peanuts were still a minor farm crop. Most southern farmers grew cotton. Then, in 1905, the South was invaded by the boll weevil, a beetle that destroyed almost half the cotton crop. During the next decade, many southern farmers were forced to leave their farms.

Fortunately, an African American scientist named George Washington Carver persuaded many farmers that they should grow peanuts. He told them that growing peanuts was easy and that peanuts also improved the soil. Today, peanuts are the tenth most valuable food crop produced in America.

The boll weevil is an insect about 1/4 inch long. It destroys cotton by laying eggs in the cotton bolls, which prevents cotton fiber from growing.

normal cotton boll

cotton boll destroyed by weevil

George Washington Carver visited farmers and wrote bulletins telling farmers about peanuts. In 1925 he gathered his peanut knowledge into a book, *How to Grow Peanuts and 105 Uses for Human Consumption.*

In the late 1800s, street vendors sold hot roasted peanuts in cities across America.

In the early 1900s, many inventions advanced peanut culture. One of the simplest machines brought peanuts to everyone for a penny . . . the peanut vending machine.

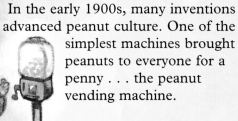

In past centuries, peanuts had to be picked from their pegs by hand. In 1905 Finton Ferguson and Jesse Benthall patented a machine that could pick peanuts automatically. This made it much easier for farmers to harvest peanuts.

peg

With peanuts now easier to pick and sell, many peanut companies sprang up. The most famous was the Planters Peanut Company, which was founded by Amadeo Obici. In 1916 he hosted a contest among school children to design the Planters mascot . . . Mr. Peanut.

Mr. Peanut ® Nabisco, Inc.

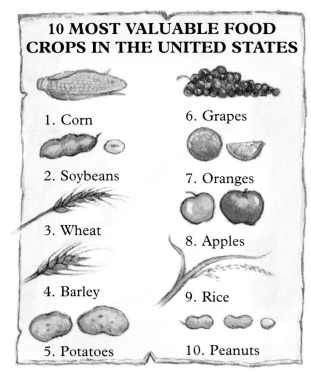

10 MOST VALUABLE FOOD CROPS IN THE UNITED STATES

1. Corn
2. Soybeans
3. Wheat
4. Barley
5. Potatoes
6. Grapes
7. Oranges
8. Apples
9. Rice
10. Peanuts

1994 National Agricultural Statistics Services

dyes

peanut recipes

printer's ink

instant coffee

peanut milk

peanut cheese

peanut flour

wood stain

axle grease

shampoo

linoleum

medicinal oils

The Peanut Wizard

George Washington Carver discovered more than three hundred uses for the peanut. Some people refer to him as "the father of the peanut industry," others respectfully call him "the peanut wizard."

For almost fifty years he conducted experiments on peanuts and other plants at the Tuskegee Institute. He promoted the peanut throughout the South and even spoke to Congress about peanuts. Known for his brilliant ideas and energetic personality, Mr. Carver helped make the peanut one of America's most popular crops.

Professor Carver found new uses for the peanut by separating its most basic parts such as its protein and oil. He would test these many ways until he discovered something new.

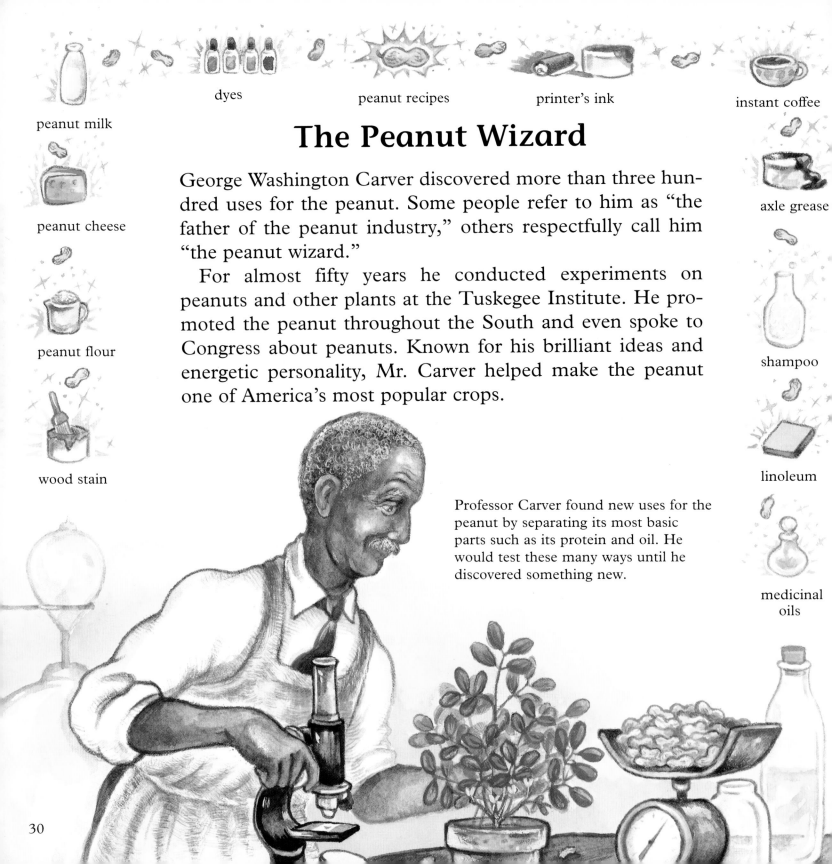

George Washington Carver was born during the last days of slavery, about 1864, near Diamond, Missouri. As a small child, George nursed weak plants back to good health, so his neighbors called him the "Plant Doctor."

When he was about twelve years old, he left home to pursue an education.

In 1894 he graduated from Iowa State College, where he studied agriculture.

He studied art and piano at Simpson College in Iowa in 1891. Two years later, one of his flower paintings won honorable mention at a World's Fair.

In 1886 George homesteaded in western Kansas. He lived in a sod house and planted his own crops.

In 1896 George Carver became director of the Agriculture Department at the Tuskegee Institute. Founded by Booker T. Washington, the school educated many African Americans.

In 1921 George Carver spoke to Congress about peanuts and other crops. His original ten-minute speech lasted longer than an hour and a half, delighting everyone. In 1953, ten years after he died, George Carver's birthplace became a national monument.

The peanut has come a long way since its humble beginnings. Today, it is a valued food and oil crop in North America, South America, Europe, Africa, Asia, and Australia. But the story doesn't end here. Every summer, billions of goobers sprout from pegs, dive into the soil, and add new chapters to the life and times of the peanut.

In January 1977, peanuts reached new heights when Jimmy Carter, a peanut farmer from Plains, Georgia, became the thirty-ninth president of the United States of America. His Georgia friends who supported him were nicknamed the peanut brigade.

Peanuts and baseball have been together for more than one hundred years. In 1908, peanuts were featured in the classic song "Take Me Out to the Ball Game." Today, vendors at some of the big ballparks sell more than one million bags of peanuts each summer.

March isn't just a leprechaun's favorite month. It is also national peanut month, and March 1 is "Peanut Butter Lover's Day."